WIRRAL
RAILWAYS
John Evans

AMBERLEY

First published 2024

Amberley Publishing
The Hill, Stroud
Gloucestershire, GL5 4EP

www.amberley-books.com

Copyright © John Evans, 2024

The right of John Evans to be identified as the Author of this work has been asserted in accordance with the Copyrights, Designs and Patents Act 1988.

ISBN 978 1 3981 1972 7 (print)
ISBN 978 1 3981 1973 4 (ebook)

All rights reserved. No part of this book may be reprinted or reproduced or utilised in any form or by any electronic, mechanical or other means, now known or hereafter invented, including photocopying and recording, or in any information storage or retrieval system, without the permission in writing from the Publishers.

British Library Cataloguing in Publication Data.
A catalogue record for this book is available from the British Library.

Typeset in 10pt on 13pt Celeste.
Typesetting by Amberley Publishing.
Printed in the UK.

Introduction

Imagine a peninsula only 15 miles long and 7 miles wide. It has nearly seventy railway stations, a line that takes trippers to the seaside and a town with more main-line stations than any other in the UK. It was the destination of the first ever electrified railway in Britain and has trains that are unique. It also hosted railways run by three of the Big Four: the LMS, Great Western and LNER all had depots and stations there.

We are looking at Wirral, home of commuters to Liverpool and Chester and a sweet spot in north-west England. It enjoys a breezy but sunny climate and a country charm with red sandstone buildings. Yet the shipyard at Birkenhead is seriously heavy industry. Run by Cammell Laird, it has built hundreds of ships for the Royal Navy and the merchant navy, including destroyers, submarines, ferries, cruisers, battleships and aircraft carriers. From the point of view of railways, it means that Birkenhead has always been furiously busy with passengers and (until fairly recently) freight. Our town with the most stations is also Birkenhead, which currently has seven, but used to have three more. And these are proper stations, not tram stops.

Over the years, and especially in the 1980s, a number of publications appeared detailing the complex history of the railways serving Wirral, mostly excellent and published locally. But they had monochrome photos and took the story up to 1980 or so. There have been so many changes that maybe it's time to look at the last fifty years in colour and to reflect on the network today.

Wirral locals will take for granted the frequent electric trains running from Liverpool underneath the Mersey and on to New Brighton, West Kirby, Chester and – with a change to diesels at Bidston – Wrexham. Yet I cannot help comparing those sixty-five plus stations in Wirral, an area of just 105 square miles, with my own home county of Northamptonshire. It is nine times larger, yet has only six stations. The purpose of this book is to tell the recent story of Wirral's railways in pictures, to depict every remaining station between 1978 and 2024 and to capture visually the huge changes – and some elements that have stayed the same – in colour.

As always, I am very grateful to those photographers who have helped me by filling the gaps. Despite the fact that I married a Wirral girl in the 1980s and have been making a nuisance of myself with my camera there ever since, I could not record it all. Those photos not credited were taken by me and the illustrations follow the sequence of the lines as detailed in the text.

Lastly, this book is dedicated to my late mother-in-law, Irene Whitehurst MBE, without whom (through her daughter) I might have remained in cheerful ignorance of all that happens when you delve beneath the Mersey.

John Evans MA
Luddenden, West Yorkshire

An Electric Transformation

'So where exactly is the Wirral Peninsula?' If you had asked me that when I was a fickle youth I would probably have muttered something unconvincing about Liverpool and tried to change the subject. Little did I know that it would eventually become a very important part of my life.

My first appearance at West Kirby, a seaside resort on the north-west tip of the Wirral, was a mixed blessing. On the one hand, in trepidation I was to meet the prospective in-laws; on the other, equally important, I needed to search out the local railway station. This turned out to be right in the centre of town and was a curiously attractive red-brick building complete with a modest clock tower. So I wandered onto the island platform to discover a businesslike three-car electric multiple unit, which according to my Ian Allan spotters' book had been constructed in either 1956 or 1938. Three sizeable front windows gave an austere external appearance, but the faded moquette seats and vintage details such as individual lampshades, straps for standing passengers and string-backed luggage racks made a ride on one of these a priority. Squeezed between the lonely coast of north Wales and the once frenetic Mersey riverside of Liverpool, the Wirral Peninsula and its coastal railway made a stressful weekend rather more palatable.

Our first dilemma is to decide what exactly defines Wirral geographically. Obviously, the northern perimeter is the sea – Liverpool Bay. To the west we have the River Dee and on the east the River Mersey. The problem comes when seeking to draw a southern boundary. Various books seem to define this as basically leading from around Burton Point on the Dee to Ellesmere Port – the physical area where the peninsula meets Cheshire. Other brave souls online give various more exact answers, ranging from specific roads to geographical features. 'Make your own mind up,' said my wife helpfully! So I have and I propose to follow my railway instincts and include a few out-of-Wirral stations; but for the most part we will be examining the electrified lines that run along the north and east boundaries.

Years ago, Wirral was a serene hideaway in north-west England, a pretty rural landscape with even prettier villages. It was mainly farming land, with hills on the Dee coast and views of water never far away. Ferries have been plying across the Mersey for 900 years and Wirral has always been strongly influenced by Liverpool, as it is today.

But the Industrial Revolution was also a key point in Wirral's development as it led to the Cammell Laird shipyards at Birkenhead and the rapid expansion of Liverpool. The peninsula soon became the home of commuters shuttling under the Mersey, but there was a snag. The trains, headed by very powerful steam engines, were an environmental disaster for passengers, who soon returned to the slower but healthier ferries. Help was at hand in the shape of George Westinghouse, a man with a product to sell and on the lookout for an ideal location. His eIectric-powered trains, technically impossible when the line was built, looked like the answer to the Mersey Railway's problems and the railway was quickly re-laid with four-rail electrification, reopening in 1902. It was an immediate success, so in the 1930s the LMS decided to extend the electrification from Birkenhead to West Kirby and New Brighton, both these towns also offering seaside leisure for local people and residents of Liverpool.

The fact that Wirral had only four railway lines sounds like things might be a trifle boring for the railway enthusiast. Not so. Of the four lines, one was a rural pre-Beeching casualty, another was a pioneer of electrification, a third for many years ran express services to London Paddington and the fourth survives today as an enchanting branch line where the ticket collector/guard has in quieter moments revealed fascinating local tourist information while displaying genuine enthusiasm and delight. You can't say that about too many British railways these days. So much has changed in the past fifty years. The freight trains, the old LMS multiple units and semaphore signals have all gone. But under current operator Merseyrail the passenger trains are extremely busy, stations have been modernised and the third generation of electric trains is now in service.

Birkenhead to Chester

The railway running along the east (the Mersey) coast was the first to penetrate the Wirral Peninsula. In 1840 a line between Birkenhead and Chester was opened, much of it in Wirral and, happily, it survives – indeed, prospers – today. Not too long after the line was opened, through trains between Birkenhead and London Paddington began running. By then, the docks at Birkenhead were already established. The railway eventually became a kind of northern main-line outpost of the Great Western Railway through mergers and acquisitions, although it was actually a joint Great Western/LMS line and in later years a mix of locomotives and stock was common.

The northern terminus of the line was initially at Birkenhead Grange Lane, but soon it was extended to Monks Ferry and later still to a new imposing station at Woodside, with its impressive overall roof, where passengers could board ferries across the Mersey to Liverpool. In the early twentieth century increasing traffic led to the lines being widened from two to four tracks from Ledsham northwards. As such it became a busy and important main line right to the end of the days of steam in 1967 when long-distance express trains were withdrawn. In the same year, the grand but faded Birkenhead Woodside station was closed and Rock Ferry became a new focal point, where passengers changed for the electrified Merseyrail system. Having been in decline for some years, the Birkenhead to Chester line then began a revival, but not as a main line. Instead, electrification of the route started in stages, not being completed from Birkenhead to Chester until 1993. It is now a very busy route with frequent trains (every fifteen minutes) which are well used from early in the morning to around midnight.

We commence our photo narrative of Wirral railways at Hamilton Square, the first station out of Liverpool after passing beneath the River Mersey. The pioneer 507 unit, No. 507001, repainted in classic BR livery, is seen at Hamilton Square on 2 February 2024.

A sad view of Class 503 EMU vehicles of the LMS 1938-built batch, dumped in Birkenhead Docks, Cavendish Road Yard, on 20 October 1984. The vehicles had been set alight to make their processing easier, and were awaiting final breaking up. (Roy Burt, copyright Gordon Edgar)

Another view of withdrawn Class 503s awaiting scrapping at Birkenhead Cavendish Road in October 1984. In the background is the Ocean Flour Mills building, a reminder of a once very important industry in the town. (Roy Burt, copyright Gordon Edgar)

Arriving at Birkenhead Central is No. 508122 on 5 April 1986. The sidings on the left led to the carriage shed.

The old Mersey Railway offices at Birkenhead Central retain the company's name. This intriguing railway under the river somehow managed to miss being part of the LMS at the 1923 grouping, and remained independent until nationalisation in 1948.

The Mersey Railway headquarters buildings are on the left as No. 777009 arrives on a damp December day in 2023.

No. 508117 stabled at Birkenhead Central depot on 14 November 1991 with a red flag denoting it should not be moved. The building still stands in poor condition. (Dave Sallery)

The remains of Monks Ferry station after demolition. Once the northern terminus of the Chester & Birkenhead Railway, it enabled access to Liverpool via a ferry service. It had three platforms and a turntable, but was eventually replaced by the much grander Birkenhead Woodside terminus. It was demolished in the 1960s. (Birkenhead Reference Library)

Birkenhead Town was another Merseyside station that was served by ferries, but it was not very convenient for the centre of the town. This was the final station before the Woodside terminus. Mollington Street engine sheds were nearby. Closed in 1945, the station site is near the entrance to Queensway Tunnel. (Birkenhead Reference Library)

Underneath the arches. This is the scene at Birkenhead Woodside on the evening of 4 March 1967, the last weekend of through trains from Birkenhead to Paddington, after arrival of the two specials hauled by Castle locomotives. (Keith Holt/KDH Archive)

On 4 March 1967 two specials were run to mark the last through trains from Paddington to Birkenhead, both headed by Castle Class 4-6-0s in preservation. Here is the first of them passing through Fenny Compton with No. 7029 *Clun Castle* in charge.

13

Type 2 diesel No. 24054 awaits its next duty at Mollington Street depot on 19 April 1976. At the time pairs of these diesels were hauling the iron ore service from Birkenhead ore terminal to Shotton steelworks. (John Woolley)

Birkenhead MPD on 3 April 1976. The depot opened in 1878 and was jointly owned by the Great Western and London & North Western Railways, who each built an eight-road shed. In 1951, the LNWR shed (on the left) was reduced to four roads to allow construction of a two-road straight shed. The depot closed to steam in November 1967 and completely in November 1985. (John Woolley)

Seen in the old Birkenhead 8H steam shed, the 508 had arrived from the Southern Region (note Strawberry Hill on the front!) and the 503 was waiting to be scrapped. (Dave Sallery)

Green Lane station is the start of a steep climb to Rock Ferry. The station was built in 1886 as a terminus for Mersey Railway trains from Liverpool, although these later continued on to Rock Ferry. Recently it has been much improved, but it retains an old-world charm.

No. 777010 waits to leave Green Lane, showing the unusual roof offering cover only on the Down platform.

On the west side of Rock Ferry station, the old terminus for Liverpool electric trains still exists, now used mainly for the stabling of stock and early morning services. There were formerly three tracks here. A freight line used to connect with Birkenhead Docks, but it was removed in 1993. In steam days, express trains stopped here on their way from Woodside to pick up passengers from the electrified services.

The signal box at Rock Ferry on 19 July 1976. It was built by British Railways in 1957 and closed on 18 September 1994 when the area was re-signalled. (John Woolley)

No. 507003 was a regular performer on the Chester line during the winter of 2023/24 and is seen here at Rock Ferry. This large station gives an indication of its former importance.

Trains passing at Rock Ferry in early 2024. Once there were six platforms – the space on the right had platforms for two more tracks.

We continue south to our next station, Bebington. Here we find a charming booking office with gables. Others on the line have similar architecture and survive to this day.

18

No. 507003 calls at Bebington with a southbound service in 2023. This used to be a very spacious station with platforms serving four tracks.

A Class 777 unit enters our next call, Port Sunlight, which was built as a 'model village' and is now a tourist attraction. There was once a siding here, but today the station is very basic and uninteresting.

On to Spital, which has a sturdy, elegant old footbridge with ramps to the platforms. No. 777019 calls in November 2023.

Spital's booking office is on the bridge and shares an architectural familiarity with other stations on the line. Sadly the Venetian-style platform buildings were swept away many years ago, but the windows and door retain a classic elegance.

Bromborough Rake was a new station opened in 1985 when electrification to Hooton was launched. It's a fairly basic station, shown here with a Class 777 arriving from Liverpool.

Very similar to Spital, Bromborough was one of the original stations on the line. Only a mile from Bromborough Rake, it is a very attractive location retaining its classic station buildings and canopies.

Not much business for this 507 unit arriving at Bromborough in 2022. This is an original Chester & Birkenhead Railway building.

Opened in 1995, Eastham Rake station's design was influenced by local residents. Note the wooden panels between the brickwork and the modern waiting shelter.

In the mid-1980s when the Pacers were new, No. 142041 in Regional Railways livery is waiting to leave Hooton for Chester, before the lines to Chester were electrified. This is the most important intermediate station south of Birkenhead.

Once Hooton was on a four-track main line from Birkenhead to Chester, and boasted through trains to London Paddington. Today the line has just two (busy) tracks, but a few rails from the old lifted lines still exist. Here a Merseyrail unit (No. 507012) for Birkenhead uses the electrified lines, while remnants of the station's former glory can be seen on the left of the picture.

23

Hooton on 13 January 2024, with No. 777017 (right) stationary after having disgraced itself by failing on a southbound working and No. 777011 arriving to rescue the stranded passengers.

Hooton station is beneath us as we look south towards the junction for Ellesmere Port, while the Wirral Way footpath to West Kirby curves to the right. (Martin Macdonald)

Happily, Hooton's station building with its curious steeple survives and is manned by an enthusiastic team of booking office staff.

North of Hooton, a Class 507 passes crossovers and industrial buildings with the old four-track formation clearly visible to the right.

Hooton station showing its elegant old footbridge on 1 July 1993 with a Metro-Cammell diesel unit on a Chester service before electrification. (Dave Sallery)

Most stations on Merseyrail now benefit from lifts, this one being at Hooton – we've lost the elegant footbridge but retained a touch of its design. Two 508 units meet, with less than a year's life before their withdrawal.

A Class 507 unit awaits departure from Capenhurst on a gloomy afternoon in December 2023. We have now left Wirral.

No. 777020 leaves Bache for Chester in 2024. On the right is a large supermarket, once the site of a coal yard. The station was opened in 1984 and replaced the less convenient halt at Upton-by-Chester, which was closed at the same time.

At the time of this picture, 13 February 1967, Chester had two stations, the other one being Chester Northgate. Here we see the splendid Italianate frontage of General station with a row of pristine taxis and various cars of the period. This station was opened in 1848.

Express services to Paddington latterly started their journey from Birkenhead behind tank engines. Here No. 42587, a Birkenhead engine, has just hauled a London service as far as Chester General. She is seen here awaiting servicing at Chester shed on 13 February 1967.

28

On 14 November 2021, No. 508126 stands at Chester having arrived from Liverpool.

A visitor from Patricroft shed is seen at Chester in the shape of Standard Class 5 4-6-0 No. 73071. Chester shed was busy right up to closure in June 1967 and on this day, 13 February 1967, every engine on shed was steam.

Hooton to Helsby

Helsby is well outside the Wirral Peninsula, but this line is within Wirral for half of its route – the busiest half. The railway here was built mainly for freight use and opened in 1863. Passenger services have been rather haphazard over the years, but Stanlow oil refinery, which dominates the local landscape, was a key driver of revenue for the railways over many years before pipelines took over.

These days passenger trains run from Liverpool to Ellesmere Port, a line electrified by British Rail in 1994, branching off the Chester line at Hooton. The train service further east between Ellesmere Port and Helsby is very limited – just two a day each way at peak hours. The line east of Ellesmere Port is not electrified.

This is Overpool, a very basic station which was not opened until 1988. Here a Class 777 waits to depart for Hooton and Liverpool in February 2024.

West of Little Sutton, our next halt, the tracks are virtually straight all the way to Helsby. No. 777010 is on the last stretch of straight track before reaching Little Sutton.

Little Sutton has an attractive Grade II listed booking hall which became part of the Merseyrail network in 1994. Ellesmere Port is similar.

31

An evening service for Liverpool waits at Ellesmere Port's attractive station, the end of the electrified system from Liverpool. The rusty rails on the right are used by occasional trains to Warrington Bank Quay.

The National Boat Museum at Ellesmere Port – a must for transport enthusiasts or anyone wanting a great day out.

A view from Ellesmere Port station towards Helsby. Notice the distant signal box and semaphore signals, the non-electrified line on the right (used by diesel multiple units) and the end of the third rail just ahead of the train.

A veteran Metro-Cammell DMU nicely captured at Ellesmere Port on 16 April 1991 working a Chester to Helsby service before electrification. (Dave Sallery)

Hooton to West Kirby

Wirral of old was dominated by scattered villages and depended on agriculture and fishing. Yet despite this, as early as 1847 there were plans to build a railway from the very attractive Deeside village of Parkgate to Hooton, on the Birkenhead to Chester line. Back in those days Parkgate was an important port with a regular packet service to Dublin, but the Dee silted up and Parkgate lost much of its trade. The railway wasn't opened until 1866 and for twenty years it was a fairly short branch. But in 1886 it was extended to the bustling seaside town of West Kirby, with its three intriguing islands, like something out of a *Famous Five* book. The railway provided a link to the Woodside ferries for businessmen travelling to Liverpool and also holiday traffic from Chester, Birkenhead and Liverpool to West Kirby's fine beaches and elegant Victorian streets.

After the direct Wirral Railway line to West Kirby from Birkenhead opened, the Hooton line became increasingly irrelevant. Passenger trains ceased in 1956 and six years later the last freight train ran. And that might have been that, but for the vision of some locals and the financial support of Cheshire County Council and the Countryside Commission. Work started in the late 1960s to convert the overgrown trackbed into a very popular 12-mile footpath known as the Wirral Way, part of the Wirral Country Park, Britain's first designated 'country park', which has provided inspiration for other similar schemes.

A quick diversion. Hadlow Road station, near Willaston, on the old Hooton to West Kirby route, has been preserved with careful attention to detail. It is the only complete station to remain on the line, although there are platforms and remnants of others. Hadlow Road once had a small goods yard.

From under-used rural branch line to popular footpath, the Wirral Way is a charming 12-mile walk with some wonderful views. If the Hooton to West Kirby railway had to disappear, this is the way to retain its memory!

West Kirby Joint station, which was the end of the line to Hooton. On the far left an electric unit can be seen in the main station as a Stanier 2-6-2T locomotive prepares to leave. (H. Leadbetter, Ian Anderson & Jon Penn archives)

Birkenhead to West Kirby and New Brighton

When the first railway along the north coast of Wirral was opened in 1866, it ran through sparsely populated land and lost money to the extent that its owner, the Hoylake Railway, had to close part of the line because a stretch of land had not been paid for. An inauspicious start maybe, but eventually rescue came and then the rails were extended to West Kirby. Connections to the docks came shortly afterwards, as did a branch from Birkenhead Park to New Brighton, a growing seaside resort popular with Liverpudlians seeking a day out. The town at one time boasted a tower taller than that at Blackpool, but it went rusty while closed during the First World War and was dismantled soon after. A long-planned link to the town of Seacombe was eventually opened in 1895. This gave more access to Mersey ferries and also served a thriving area. After a number of takeovers and name changes, the whole conglomerate settled down as the Wirral Railway. At Birkenhead Park, there was a connection with the Mersey Railway, which proceeded through the tunnel into Liverpool. Seacombe station was a fairly rudimentary low-cost affair, and when the line closed in 1963 much of the route was converted into an access road for the new Kingsway Tunnel.

At Hamilton Square two mighty pumping stations made a desperate attempt at cleaning the smokey air in the tunnel. An underground station beneath Liverpool Central became the city destination. Once Westinghouse's electrification plans had been completed, a very frequent service ensued including trains to Birkenhead Park where passengers to West Kirby and New Brighton had to change trains until these lines were also electrified in 1938.

We now return to Birkenhead to set out for West Kirby. Deep underground, a grubby Class 507, No. 507032, approaching the end of its life, stands at Hamilton Square station in November 2023. It is heading towards Liverpool.

Large murals were painted at Hamilton Square station to celebrate the Mersey Railway centenary in 1986.

A tall tower was built at Hamilton Square to house three 100-passenger hydraulic lifts, which allowed people to descend to the station platforms well underground. It was once driven by a grasshopper beam engine. Later, smaller lifts driven by electric motors were installed.

Conway Park was built in 1998 to provide a station closer to the centre of Birkenhead. Here, No. 507010 emerges from the underground section into the dappled daylight.

Conway Park was excavated by digging down from ground level, so it has daylight above. A warning sign tells the less energetic that there are ninety-nine steps to ground level (but there are also lifts). No. 507023 comes to a halt

The approach to our next station, Birkenhead Park, on 4 April 2015, with No. 508128 in action on a lovely spring day.

39

A year later, on 14 April 2016, the same unit, No. 508128, stands at Birkenhead Park station, once the change point between electrified services to Liverpool and steam trains to West Kirby. The building on the bridge replaced an earlier one destroyed during the war. There was once another platform to the left of the train.

Arriving at Birkenhead North in steady rain is the now-preserved pioneer Class 507 No. 507001, repainted in its original colours, with a West Kirby train on 2 February 2024.

40

A busy scene at Birkenhead North station on 23 August 1982 – a Class 503 unit is arriving on a train for West Kirby. A newly refurbished set is on the left.

No. 507128 stands at Birkenhead North on a pleasant day in July 2020.

41

No. 507020 sports an ornate variation of the normal Merseyrail livery when seen at Birkenhead North on 13 May 2015.

Another view at Birkenhead North on 25 April 1984 with a Class 503 waiting to leave for West Kirby, while its replacement, a new Class 508, No. 508015, is stabled on the right. Notice the neat little signal box.

This distinctive green bridge allows the Chester Road to pass over the entrance to Birkenhead North station. The line to the left gives access to the traction maintenance depot behind us as No. 507001 arrives from Liverpool in 2016.

Occasionally a unit is turned out in very different colours. Waiting at Birkenhead North is Class 507 unit No. 507002 in all-over Liverpool Hope University colours on 6 October 2017.

Two Mersey multiple units were returned to their original colours to celebrate 100 years of the Mersey Railway in April 1986. The restored two-car unit is at Birkenhead North on centenary day, with four tracks running through the station. This station replaced an earlier one west of Bridge Road.

No. 508108 arrives from New Brighton at Birkenhead North with Wallasey Bridge Road in the background. Beyond the bridge is the traction maintenance depot, on the site of a one-time steam shed which was not required following electrification.

By 2017 the Class 507s were the oldest passenger trains in regular use in the UK apart from Class 313. They were delivered new to the Wirral line more than thirty years earlier. Here No. 507028 arrives at Birkenhead North on a dull 14 November 2017 with a service from New Brighton.

A Liverpool-bound train in 2024 rounds the curve by the traction maintenance depot to reach the lesser-used platform 1 at Birkenhead North.

45

Changes are afoot on the Wirral line in 1984. We see a newly delivered – from the Southern Region – Class 508, still bearing the destination of its last journey, to Effingham Junction in Surrey. An old Class 503 unit stands in the platform at Birkenhead North on its way to Liverpool Central. Before long the 507 and 508 units will be running all services on the Wirral and Mersey lines.

'Beatles Story' liveried Merseyrail unit No. 508111 approaches Birkenhead North working with a service from West Kirby on 20 December 2016. No. 507011 in standard colours is behind. (John Woolley)

A single passenger leaves Birkenhead North station on a miserable afternoon just three days before Christmas in 2022. This was the last full year for all the 508s, but this 507 was still working in early 2024.

Electro diesels Nos 73004 and 73005 at Birkenhead North, 5 March 1994, with No. 508140 passing on a service to Liverpool. The Class 73s were used as shunters, although No. 73004 was just a source of spares. (Dave Sallery)

47

As part of the Mersey Railway centenary celebrations, a variety of guest locomotives was displayed at Birkenhead North, including the preserved Type 4 diesel No. D200.

The usual shunter at Birkenhead in 1986 was 0-6-0 No. 03170, seen here at Birkenhead North depot. It was one of the last survivors of its class and is now preserved.

Departmental unit No. DB 975179, a conversion to battery power of a Watford electric unit, stands at Birkenhead North depot alongside a Class 503 Wirral unit in March 1976. Both units are painted plain BR blue. (Roy Burt, copyright Gordon Edgar)

A former multiple unit used on Euston to Watford electric services is in the unlikely surroundings of Birkenhead North depot in the summer of 1986. It was here having been converted to a Sandite and de-icing unit, and survived until 2002 when it was preserved.

An open day at the nerve centre of Merseyrail. Inside the maintenance depot at Birkenhead North in 1986 a multiple unit is being repainted, with a pair of Class 03 shunters at the back of the workshop.

The two-car unit on centenary day passes Birkenhead North depot in April 1986.

Class 03 shunter No. 03189 stabled at Birkenhead North depot on 20 October 1984 in the company of a Class 503. The 03 was withdrawn from service at Birkenhead on 16 March 1986, following the return from Swindon Works of sister loco No. 03170. It was saved for preservation and today is to be found at the Ribble Steam Railway at Preston. (Roy Burt, copyright Gordon Edgar)

Sprinter No. 150254 comes off the Wrexham branch to join the West Kirby to Liverpool line at Bidston station in January 2020. In the far distance, a train from West Kirby is identifiable by its headlights. In the old days this was a lonely interchange station with little local custom.

51

A meeting of trains at Bidston station on the Wirral line. A Class 503 electric multiple unit is ready to head under the M53 motorway towards West Kirby, while a diesel multiple unit has arrived from Wrexham. Taken on 25 April 1984, today everywhere is much more overgrown.

Thirty years later, a Class 508 unit, No. 508108, sets off from Bidston for Liverpool. This is the interchange station for regular diesel trains to Wrexham, which now all use platform 2 and disappear round the curve to the left. The M53 motorway to Birkenhead is in the background. The line to Leasowe here is dead straight and you can stand on the platform and watch trains for several minutes as they approach.

In 1986, the M53 looks very quiet in the background as No. 508103 arrives at Bidston. The standard LMS Dee Junction signal box was swept away in the 1990s re-signalling programme. The area on the left was once busy with sidings leading to Bidston locomotive depot.

An evening view at Bidston station. Class 150 No. 150253 provides the service to Wrexham connecting with the Wirral line in June 2016.

53

On 24 April 1984 the old order at Bidston still rules, with a three-car Class 503 unit passing under the A554 road on its way to West Kirby. The line curving left goes to New Brighton (and formerly to Seacombe), while Birkenhead Docks are in the background. The road on the right used to be sidings which led to Bidston loco depot. It was a bleak and isolated location on a miserable winter evening.

The restored two-car unit approaches Bidston in 1986 with the old curve to New Brighton still in place.

The service for Wrexham on the left, with a Class 150 Sprinter, has little time to prepare for its return journey as it has to fit in with the busy service to West Kirby. No. 507008 has just supplied some passengers for the Borderlands line, as the Wrexham branch is now called.

Bidston Ore Dock, Birkenhead, on 27 May 1989, with Baguley Drewry 0-4-0 diesel-mechanical locos *Kathleen Nichols* and *Dorothy Lightfoot* mothballed at Rea Bulk Handling. No further rail traffic came for them and they were eventually sold on for further use elsewhere. This excellent photograph shows the dock where ore was imported for John Summers steelworks at Shotton. (Gordon Edgar)

A new Transport for Wales two-car unit, No. 197018, brings a splash of colour to Bidston in December 2023 as it waits to leave for Wrexham.

Dee Junction at Bidston, with No. 507030 approaching from West Kirby. The rusty rails are the little-used Up line towards Birkenhead, because regular services now arrive and depart on the Down side.

The east junction at Bidston is very simplified compared with years ago, when there were extensive sidings on the right and a freight line to Rock Ferry. The line to the left in front of No. 507008, which is on its way to Liverpool, once led to New Brighton as part of a triangle. Now it is a short siding.

Between Bidston and Leasowe, No. 777024 traverses the dead straight tracks with Bidston station in the distance.

57

A train crosses the level crossing at Leasowe station heading for Liverpool. One of the dogs seems fascinated, but the other couldn't care less!

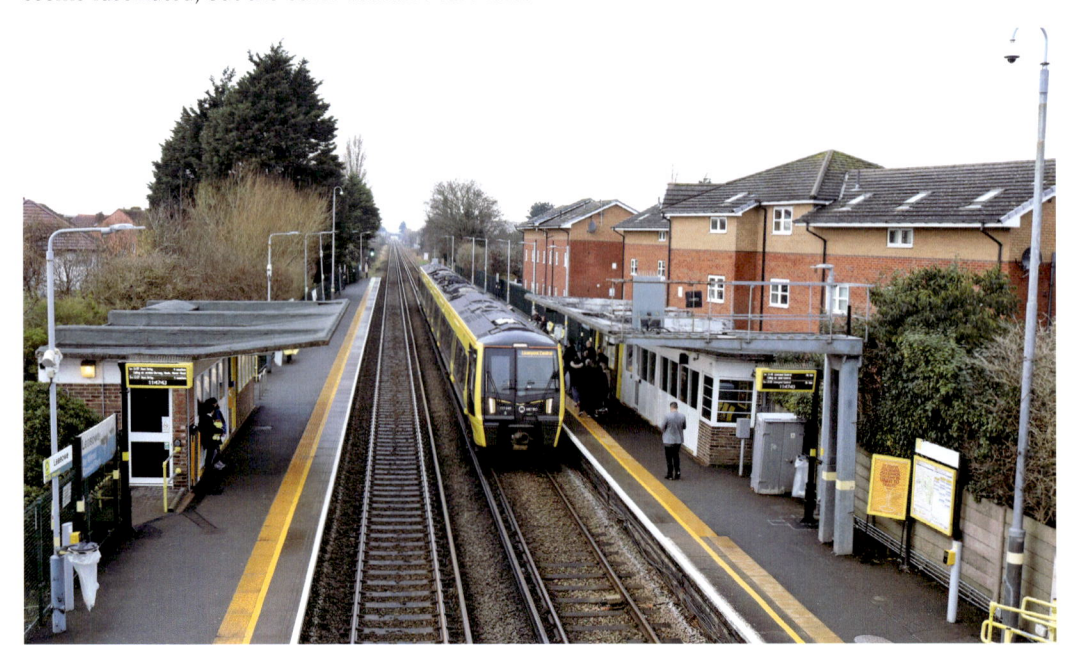

Leasowe station viewed from its footbridge, with a small ticket office of the usual LMS 1930s style hidden away in a platform building,

In July 1986, when Class 508 units were quite new, No. 508131 is heading for West Kirby on the long straight on the north coast. Semaphore signalling is still in action.

A train arrives from Liverpool at Moreton, which can be seen in the background. On the left is the Typhoo Tea factory with its stumpy chimney, rail-served until 1971. It used to make Cadbury's cakes and had its own little engine.

The spacious booking office at Moreton is on the bridge and there were once sidings here serving a brick works. Two Class 777s meet under the concrete platform roofs.

No. 508138 leaves Meols for West Kirby on 30 May 2021. The signal box here was closed in the 1930s and the station now boasts modern passenger lifts.

The West Kirby line is never far from the sea. Meols Pond (created by Braithwaite Poole to attract anglers in summer and skaters in winter) is in the foreground, with a train in the station in February 2024. (Martin Macdonald)

No. 777003 enters Meols in February 2024 with the long straight from Moreton in the background. The sea is about half a mile to the left.

61

A timeless view that shows the two-car restored unit at Manor Road amid typical Wirral housing on 5 April 1986.

In March 1985, No. 508104 runs into Manor Road, opened in 1941 to serve new housing areas at the east end of Hoylake.

A small boy watches No. 508103 arriving at Manor Road on 5 April 1986.

This picture was taken on a bright April day in 1984 and shows a Class 503 unit at Manor Road station, near Hoylake. Nearest the camera is No. M28383M. This is one of the original 1938 units, still giving good service and one of the few pre-war trains then running on BR.

This Wirral and Mersey Class 308 unit is seen leaving Manor Road station in April 1984. It was the last year that these old-timers were in charge, some of which were built by the LMS before the war, although this one dates from 1956.

No. 508128 on the gentle curve between Manor Road and Moreton in September 2011.

No. 508115 prepares to leave Hoylake eastbound on 29 July 2020. This was the original terminus of the Wirral Railway, before the extension to West Kirby. A gasworks once occupied the site on the right, where the platforms have been extended.

In 1986 the restored two-car unit stands in Hoylake station, awaiting departure for Birkenhead North. Note the oil lamp.

No. 508108 enters Hoylake with a service from Liverpool in 2020.

This was a very lucky shot on 6 April 1986. Just outside Hoylake station, the old Class 502 unit passes Class 508 No. 508113. I had my 135 mm telephoto lens installed and managed this picture with my trusty Pentax ME Super.

No. 508115 leaves Hoylake for Liverpool in 1986, when a siding was still present on the right. Once a carriage shed and maintenance depot stood here and there were also sidings on the left.

A six-car rake seen here between West Kirby and Hoylake in 1987, passing the municipal golf course. Just visible on the horizon are the Welsh hills across the River Dee.

No. 507024 arrives at Hoylake amid new trees with a Liverpool to West Kirby service in June 2020.

The restored two-car unit at Hoylake on 6 April 1986, the 100th anniversary of the Mersey Railway. It never ran on these tracks in normal service, being confined to the Southport line.

The concrete station at Hoylake is seen here on 25 April 1984. These days its architecture is much more appreciated as it represents a time and style. Today the Wirral Railway signal box, signal and crossing gates have all gone.

No. 508140 at the road crossing in Hoylake, with maintenance taking place. Compare this scene with the previous picture.

69

Another centenary shot at Hoylake, with the restored two-car unit running round in 1986.

A busy Wirral scene at Hoylake with a train (left) setting off for West Kirby, which you can see in the distance, while another approaches. Both trains are Class 503 units and the date is 25 April 1984. The Hoylake municipal golf course is on the left, with the houses of Drummond Road on the right.

A train for Liverpool Central stands at Hoylake in May 1983. The two-tone livery suited these units much better than the dreary plain BR blue they previously carried.

Sidings were once laid either side of the approach to Hoylake from the north, but by 1982 when this picture was taken, some rationalisation had taken place. An electrified storage siding seen behind the train remained at this time.

A Class 503 electric multiple unit enters West Kirby past the semaphore signals that used to protect the station. It is 7 May 1983.

In 2020, No. 507025 prepares to leave for Liverpool on a humid July day.

Between West Kirby and Hoylake, the line passes the backs of houses. Two Class 508 units are seen in August 1988.

The refurbished concrete roof at West Kirby station can clearly be seen on 28 July 2023, a perfect summer's day, as No. 507033 waits to leave for Liverpool.

73

In August 1986 two Class 508s are stabled at West Kirby. The coast is just a five-minute walk from the station, so day-trippers provide plenty of custom in the summer.

A train for Liverpool awaits departure from West Kirby in 2019, with typical modern platform clutter that has replaced the lovely old bracket signals. Every year the foliage encroaches more.

A Class 503 unit approaches West Kirby from Hoylake. The industrial building on the right used to be a bus depot. It has gone since this picture was taken on 25 April 1984 and the undergrowth is higher. The driver has already reset the destination to Liverpool Central.

A July day in 2020 sees No. 507023 leaving West Kirby, with a much simplified track layout compared with older photographs.

75

In the summer of 2020, No. 507013 is ready to leave West Kirby for Liverpool. A supermarket now stands alongside the line at this point.

The end of the Wirral line is West Kirby with its very pleasant station building. Here, unit No. 507006 in the old livery awaits departure for Liverpool on 22 March 2010. It was all so different in the 1950s, when two stations stood within a short distance. The modern roof on the left marks the site of the closed line to Hooton.

New to Wirral, ex-Southern Region unit No. 508107 is stabled at West Kirby in May 1986.

This view in July 1986 of No. 508112 leaving West Kirby offers a panoramic aspect of the station. The fire station on the left was once the site of the second West Kirby station, which served trains to Hooton. The author owns the signal box sign.

In 2014 major refurbishment was carried out on West Kirby's station canopy, seen here during the work as No. 507001 (in the old yellow/silver-grey livery) waits to leave for Liverpool.

The façade of West Kirby station includes an access road (once a taxi rank) and this attractive red-brick tower with clocks and battlements. The gables are half-timbered.

Spectacular sunsets are just one reason why people visit West Kirby, as seen here looking towards Hilbre Island. (Jane Evans)

Class 508s have now disappeared, but No. 508120 was stabled at West Kirby on 30 May 2021 in the morning sunshine.

West Kirby station in February 2024, with low tide in the Dee estuary and the Royal Liverpool golf course (top right). The old Hooton line passed beneath the road where the trees in the foreground are now. (Martin Macdonald)

Semaphore signals and the signal box installed in 1937 for electrification the following year dominate this view of a Class 503 arriving at West Kirby on 29 August 1982. This was my first ever Wirral picture!

We move to New Brighton, where the heritage Class 503 three-car unit makes a fine sight on 6 April 1986. This train used the now closed curve at Bidston to access the West Kirby line.

The restored Class 503 unit leaves New Brighton on centenary day in 1986 with semaphore signals still in place. Liverpool Bay is in the background.

81

On a chilly winter's day in 2017, No. 508141 is waiting to leave the concrete and brick surroundings of New Brighton. (John Woolley)

A busy moment at New Brighton in 1986, with one unit departing for Liverpool and another having just arrived. In the 1920s there was a through coach from here to London Euston.

A pair of Merseyrail electric units at New Brighton in 2005. (Dave Sallery)

The station buildings at Wallasey Grove Road, our first stop after New Brighton, have changed little over the years. There was once a handsome clock in the bay window and the air-conditioning unit is hardly attractive, but a traveller from the 1950s would instantly recognise this as Grove Road.

83

A quiet Sunday at Wallasey Grove Road in 2021 as No. 508103 approaches the empty platforms. The sea is in the distance.

A busy moment at Grove Road, with a Class 507 leaving for Liverpool on a sunny February day while a Class 777 waits to head north. There was another station, Warren, between here and New Brighton, but it closed in 1915 due to competition from trams.

The elegant covered footbridge at Grove Road with stone supports, of which there were quite a few like this on Wirral railways. Many have since been rebuilt or replaced.

A serene picture at Grove Road with No. 508124 moving away to New Brighton.

No. 507023 runs into Wallasey Village on a Liverpool service. The station architecture here reflects that seen all over north Wirral with simple but elegant concrete and brick buildings. Nearby Wallasey Grove Road is in the background.

No. 507020 arrives at Wallasey Village from New Brighton. Despite some modernisation over the years, this station looks basically the same as it did when built for electrification in the 1930s.

A Class 507 leaves Wallasey Village for Birkenhead North and Liverpool, crossing the dual carriageway which dominates the site of this station.

A new Class 777, No. 777028, crosses the road and enters Wallasey Village in February 2024.

Hawarden Bridge to Bidston

The last line in the Wirral Peninsula was from Bidston on the Wirral Railway's West Kirby line to Hawarden Bridge, where it linked to the route from Wrexham to Chester Northgate. This later became part of the Great Central Railway and then the LNER. Trains ran from Wrexham to Bidston, just outside Birkenhead, but were later extended to Birkenhead Park and then to Seacombe. After Seacombe station closed to passengers in 1960, services were diverted to New Brighton, but today they run into Bidston. There have been plans to electrify what is now called the Borderlands line, but it has been ruled out on cost grounds so far. In the days of steam, the route was very busy with heavy iron ore trains running from Bidston Dock to the important John Summers steelworks at Shotton. Despite the closures, there is much to explore for both tourist and enthusiast.

A Transport for Wales diesel unit, No. 197014, arrives at the very quiet station of Upton, the first after Bidston, which once had a booking hall on the bridge and proper station buildings.

Originally named Heswall Hills, today's Heswall station is approached through a housing estate which was the site of a busy freight yard. It's a stark, uninteresting station, with the most rudimentary of platform buildings.

Passengers face a good walk from the car park up to the platforms. There used to be a booking hall here, leading through a covered walkway to wooden platforms and elegant waiting rooms.

Another new face of trains on Wirral. A Vivarail rebuild of ex-London tube stock No. 230008 is seen here approaching Heswall in 2024. They've swapped Barking for Bidston and probably prefer the country air!

Running towards Heswall is No. 197044 in February 2024. Unlike the Chester and West Kirby lines, the Borderlands route has remained quite rural.

By the 1980s, the nice old station at Neston had not changed much since steam days. Here a Derby Lightweight DMU halts with a Wrexham service on 22 April 1982.

It's hard to imagine this is the same place as the previous picture. Forty-two years has seen the old buildings demolished, but the station is busy, and notice the little tower at the Aldi supermarket reflecting the design of the old station.

Trains passing at Shotton High Level on 2 June 2016. Sprinter No. 150267 is departing for Wrexham, while another Sprinter approaches on a reciprocal working. In those days Arriva ran the services.

91

The end of the line. Sprinter No. 150241 on the approach to Wrexham General from Bidston on 11 January 2020.

No. 508015 pauses at Clapham Junction with a service to Shepperton on 5 April 1980 when it was almost new. Three years later these units were transferred to Birkenhead, this one lasting until the summer of 2023. Not a bad investment. (John Woolley)

Movement of lime from the Peak District to the docks on the Mersey sent engines from Birkenhead depot to Buxton, latterly usually a Class 9F. Here No. 92120 from Birkenhead is on Buxton shed in June 1966 waiting to take a load of lime back to Merseyside, with another Birkenhead 9F behind.

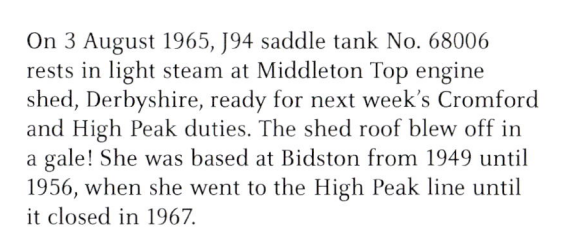

On 3 August 1965, J94 saddle tank No. 68006 rests in light steam at Middleton Top engine shed, Derbyshire, ready for next week's Cromford and High Peak duties. The shed roof blew off in a gale! She was based at Bidston from 1949 until 1956, when she went to the High Peak line until it closed in 1967.

Built by Beyer Peacock, the Wirral Railway owned three 4-4-4Ts and a fine sight they must have made running across the flat Wirral landscape. Their big bunker and tanks gave them an excellent range, but when the line was absorbed into the LMS in 1923 they were quickly withdrawn. (Birkenhead Reference Library)

Although the old Class 502 units were prohibited from using the Wirral lines as they had no front tunnel escape doors, that was not the case with the Class 507s and 508s. These ran anywhere on the Merseyrail network and here is No. 508138 in the 1990s yellow and white livery at Southport.

The old signal box sign from West Kirby. In 1995 you could buy relics from a British Rail shop near Euston and on this day (after the Wirral re-signalling) there was a big choice of local signs, so I chose this one.

We end at the beginning. This intriguing picture shows a vintage train at West Kirby Joint station. It was taken in early LMS days when trains on the Hooton line still used these classic coaches. (H. Leadbetter, Ian Anderson & Jon Penn archives)

Bibliography

Atkins, C. P., 'Mersey Railway Tank Locomotives', *Railway World, March 1976* (Shepperton: Ian Allan, 1976)

Boumphrey, I. and M., *Railway Stations of the Wirral* (Prenton: Ian Boumphrey, 2002)

Brack, A., *The Wirral* (London: Batsford, 1980)

Burnley, K., *Portrait of Wirral* (London: Robert Hale, 1981)

Gagan, J. W., *Steel Wheels to Deeside* (Birkenhead: Birkenhead Press, 1983)

Jermy, R., *A Portrait of Wirral's Railways* (Birkenhead: Countyvise, 1987)

Merseyside Railway History Group, *The Hooton to West Kirby Branch Line and the Wirral Way* (Central Library, Birkenhead, 1982)

Mitchell, V. and K. Smith, *Birkenhead to West Kirby* (Midhurst: Middleton Press, 2014)

Acknowledgements

With thanks to Abigail Garrad at Birkenhead Reference Library, the Merseyside Railway History Group, John Woolley, Gordon Edgar, Jane Evans, Dave Sallery, Martin Macdonald and Alistair Holt for their help with this book. Also to the very friendly staff on board both Borderlands and Merseyrail trains.